the Weekend Crafter®

Papermaking

Beautiful Papers and Projects to Make in a Weekend

CLAUDIA K. LEE

LARK
BOOKS

A Division of Sterling Publishing Co., Inc.
New York

ART DIRECTOR & PRODUCTION:
KATHY HOLMES

PHOTOGRAPHY:
EVAN BRACKEN

EDITOR:
KATHERINE DUNCAN AIMONE

EDITORIAL ASSISTANCE:
HEATHER SMITH

ILLUSTRATIONS:
ORRIN LUNDGREN

TEMPLATES:
MEGAN KIRBY

PRODUCTION ASSISTANCE:
HANNES CHAREN

Beverly Plummer, plant papers.

Library of Congress Cataloging-in-Publication Data

Lee, Claudia K.
    Papermaking : beautiful papers and projects to make in a weekend /
Claudia K. Lee.
        p.   cm.—(The weekend crafter)
    Includes index.
    ISBN 1-57990-194-8 (pbk.)
    1. Paper, Handmade.   2. Papermaking.   I. Title.   II. Series
TS1124.5 .L44   2001
745.54—dc21

                                        00-065503
                                             CIP

10 9 8 7 6 5 4 3 2 1

Published by Lark Books,
a division of Sterling Publishing Co., Inc.
387 Park Avenue South, New York, N.Y. 10016

© 2001, Claudia K. Lee

Distributed in Canada by Sterling Publishing,
c/o Canadian Manda Group, One Atlantic Ave., Suite 105
Toronto, Ontario, Canada M6K 3E7

Distributed in the U.K. by Guild of Master Craftsman Publications Ltd.
Castle Place, 166 High Street, Lewes, East Sussex, England BN7 1XU
Tel: (+ 44) 1273 477374
Fax: (+ 44) 1273 478606
Email: pubs@thegmcgroup.com
Web: www.gmcpublications.com

Distributed in Australia by Capricorn Link (Australia) Pty Ltd., P.O. Box
6651, Baulkham Hills, Business Centre NSW 2153, Australia

The written instructions, photographs, designs, patterns, and projects in
this volume are intended for the personal use of the reader and may be
reproduced for that purpose only. Any other use, especially commercial use,
is forbidden under law without written permission of the copyright holder.

Every effort has been made to ensure that all the information in this book is
accurate. However, due to differing conditions, tools, and individual skills,
the publisher cannot be responsible for any injuries, losses, and other
damages that may result from the use of the information in this book.

If you have questions or comments about this book, please contact:
Lark Books
50 College St.
Asheville, NC 28801
(828) 253-0467

Printed in China

ISBN 1-57990-194-8

the Weekend Crafter®

Papermaking

CONTENTS

INTRODUCTION

THE UNIVERSAL APPEAL of making paper by hand is undeniable. In the past, handmade paper was almost exclusively used as a surface for printmaking, letterpress, and painting. Today, things have changed.

All over the world, artists are inventing new ways of working with paper pulp to create art that is both exciting and innovative. Thanks to the Internet and several international paper organizations, there is an exciting ongoing dialogue between all kinds of people who love paper.

Children are making paper in school, and graphic artists are using custom-made papers to personalize their presentations. Handmade papers are commonly used for wedding invitations and birth announcements, creating collectible mementos of special occasions.

This book conveys the magic of papermaking to those of you who may not have experienced this process yet. For those of you who already love to make paper, it is my hope that this book will lead to inspiration and growth in your work. The step-by-step projects will show you how to create basic papers that can be used to make a series of decorative and functional objects. And a gallery of work by accomplished artists will inspire you and show the innovative ways in which handmade paper is being used to make fine craft and art. Try out various projects, learn the ways of papermaking, and invent your own elegant paper presentations.

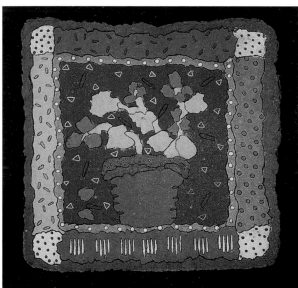

Top right: **CLAUDIA LEE**, *Bamboo Letter*, Assembled handmade paper, sand, and silkscreen, 8 x 11 inches (20.3 x 27.9 cm)

Middle left: **CLAUDIA LEE**, *Geranium*, Assembled handmade paper pieces with painted embellishment, 15 x 15 x ½ inch (38.1 x 38.1 x 1.3 cm)

Right: **CLAUDIA LEE**, Handmade/batiked paper books with Coptic binding, 4 x 4 inches (10.2 x 10.2 cm), 8 x 4 inches (20.3 x 10.2 cm), 4 x 3 inches (10.2 x 7.6 cm)

SETTING UP A PAPER STUDIO OR WORKSPACE

YOUR STUDIO CAN be as simple as the corner of a room with a plastic vat where you keep all your papermaking equipment, or it can be a room devoted to your pursuit of paper.

You'll need the following things for your workspace:

1. A well-ventilated space with good light—features which are important to your health and the enjoyment of the time you will spend working. If you intend to wax your paper (see page 20), make certain that you have adequate ventilation.

2. A convenient source of water (an outdoor hose adjacent to your workspace works best) and a place for draining vats and buckets of pulp. Since pulps are made of natural materials, it's safe to pour drained water outside, or pour it into a sink after you've thoroughly strained the pulp.

3. A table on which to work and one on which to place a drying rack. Adjust the tables so they are a good height for you. If you're a tall person, you can elevate them on cinderblocks or bricks. It's important to be comfortable while working. Pay attention to what your body's telling you.

4. Surfaces on which to dry papers, such as sheets of plastic light diffuser material, countertop material, glass, or mirror.

5. An indoor or outdoor clothesline for drying felts and cloths.

6. Storage space for pulps and equipment such as an empty plastic tub with a lid, a shelf, or some large, empty drawers.

7. A dry work and storage space, such as a table and shelves, for working and storing finished papers.

AN OVERVIEW OF BASIC PAPERMAKING EQUIPMENT

PAPERMAKING IN ITS MOST BASIC form requires a container to hold water and pulp and a frame with a screen attached to it. The following section will introduce you to basic equipment that you'll need to make paper at home. As you undertake the projects, you'll also be provided with a list that includes equipment specific to that project.

If you plan to make paper from natural plant materials (see project on pages 27 to 29), you may want to buy a pair of rubber boots and heavy cotton gloves for gathering plants in the field. A plant identifying book and a field notebook to jot down where and when you found a plant will help you identify and record plants.

To make paper from plant pulps, you'll need a large hanging scale for weighing dry plant materials. For cooking the plants, you'll need a hot plate or stove, a large stainless steel pot, a supply of soda ash and a ounce/gram scale for weighing it, and a wooden stirring rod or a length of PVC pipe.

You'll need equipment for processing materials to create pulp. You can process plant fibers by hand by beating them on a wooden board with a mallet from a papermaking supplier, a wooden baseball bat, a length of 2 x 4, or a mallet used for

Weigh out plant materials in a scale.

Equipment for cooking and straining pulp includes a stainless steel pot and stirring stick for cooking plant pulps, measuring spoons and cups for apportioning chemical additives, large and small collanders for straining pulp, and buckets for catching the water drained from pulp.

tenderizing meat. A kitchen blender works well for processing cotton linters (sheets made of cotton fibers found close to the seeds), abaca (Manila hemp), recycled pulps, and many softer plant fibers. A blender can also be used to mix materials to be used for inclusions such as marigolds and leaves (see page 17). If you can find a used blender at a thrift store, you needn't spend money on a new one.

You'll need a couple of colanders (large and small) for straining pulp out of water. A large colander from a restaurant supply house works well for straining bigger batches of pulp.

You'll also need easy-to-clean containers for holding water and pulp. Large plastic storage containers work well for this purpose, and it's handy to have an assortment of sizes on hand. These containers come with lids that make them ideal for storing papermaking equipment when they're not in use.

You can process pulp in a blender, or beat it by hand with a mallet or stick on a piece of board.

A blender can be used to process pulp or mix materials to be used as inclusions.

Plastic tubs in different sizes are used to hold water and pulp. Choose a tub that is just large enough to accomodate dipping your mould and deckle.

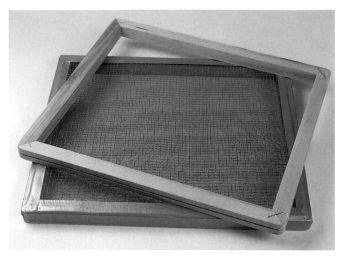

A mould and deckle comes in two parts—a screened frame for straining the pulp and a plain frame for containing the edges of the paper.

Equipment for pressing and removing water from paper include felts, non-woven interfacing, fiberglass screening, a sponge, and a drying screen. You can press papers by hand, or use a small bookbinder's press.

The size of your mould and deckle will determine the size of your paper. You can fit an embroidery hoop with fiberglass screening to make a round piece of paper.

To form sheets of paper, you'll need a mould and deckle—a simple tool made up of two frames of the same size. (See page 13 for instructions on how to make a mould and deckle.) The paper is formed on the mould—a water-resistant frame with a taut screen attached. The deckle is an open frame that sits on top of the mould and holds the pulp on the screen during the paper-making process. (You can choose to use the mould without the deckle, but the edges won't be contained.) This piece of equipment should fit comfortably into the vat or tub when you dip it. For occasional masking of the screen of your mould and deckle, you'll need duct tape or masking tape.

After you've pulled a sheet of paper out of the vat, you'll transfer it, or couch it, onto a felt. Felts are absorbent pieces of fabric that support wet sheets of paper during pressing. You can use heavy, woolen felts made specifically for papermaking (see supplier on page 79), old woolen blankets, cotton sheeting, or muslin. They should be cut larger than your mould and deckle so that the sheet of newly formed paper fits comfortably on them.

You'll need to have a large sponge for pressing and removing water from the paper before and after you transfer it to a felt. Fiberglass screening can be used to protect the paper that's been transferred to felt before you sponge it again.

A drying surface for your freshly pressed sheets can be as simple as a sheet of plastic light diffuser grid, countertop material, glass, or mirror. Any surface that is flat and water-repellent will work. Purchase these materials from a building supplier, and store them by leaning them against a wall until you need them. When you're ready to dry papers, spread the drying surfaces out on a tabletop or between sawhorses in a well-ventilated area before laying out your papers.

If you take up papermaking as a serious hobby, a small bookbinding press is a nice piece of equipment to own. You can purchase one from a papermaking supply company (see page 79). After you've transferred wet paper to a felt, place another felt on top of it before pressing it. A press applies even pressure, and removes the water from the paper efficiently. The pressure also serves to bond the fibers of the paper.

PULPS TO MAKE AT HOME

SEVERAL PULPS CAN be made easily and inexpensively at home by beating them or processing them in a blender.

PLANT PULPS

Plant pulps are made from cooked and processed natural materials such as straw and corn husks as well as leaves such as daffodil, lily, yucca, iris, pineapple, and bamboo. For specific step-by-step instructions on how to make paper from plants, see the project on pages 27 to 29.

PURCHASED PULPS

You can make beautiful sheets of paper from purchased fibers such as cotton linter (sheets made from fibers found close to the seeds) and

Plant pulps are made from natural materials such as straw, corn husks, iris, and bamboo.

Plant papers have beautiful, textured surfaces.

Sheets of paper made with cotton linter and abaca

abaca (a banana leaf fiber). These fibers come in sheet form and can be purchased from a papermaking supplier. Cotton linter is made up of short fibers that make a soft paper, and works well for casting forms as well as making sheets of paper. Abaca is also known as Manila hemp and makes a thinner, crisper sheet than cotton linter. To use either of these fibers, soak them in water before tearing them into small, quarter-sized pieces to process with water in a blender. You can also mix the two pulps together in the same vat to create a nice paper.

RECYCLED PULPS

Recycled pulp can be made from all that stuff that you usually get rid of— junk mail, newspapers, and copy machine papers. (See the projects on pages 30 to 33 for specific instructions.) If you recycle mail, be sure to remove the staples and plastic windows on envelopes before making it into pulp. Glossy magazines don't recycle well since they contain clay filler, which will give your papers a dusty surface.

To recycle papers into pulp, cut or tear the papers into small, quarter-size pieces, and soak them until they're saturated. Process them with water in a blender. The amount of time that you spend processing the paper will determine the look of the paper. Experiment with this factor to create different kinds of pulp. A short turn in the blender will create pulp with larger pieces of the recycled materials in the finished sheet, and a longer run will produce a more homogenized paper.

ADDING SIZING TO PULPS

By adding sizing to your pulp, you can reduce the absorbency of your paper so that it can be used for writing or painting. Sizing also provides protection for the surface of the paper.

As a rule, add 3 tablespoons (45 mL) of liquid sizing per pound (454 g) of dry fibers to your pulp (weigh the fibers before soaking them or beating them). Dilute the sizing in water before adding it to your pulp.

Sizing can be purchased from a papermaking supplier (see page 79). Liquid sizing has a shelf life of several months and needs refrigeration. Follow the directions that come with the sizing that you purchase.

Recycle common sewing patterns into elegant papers.

HOW TO MAKE A MOULD AND DECKLE

Build a basic mould and deckle for pulling sheets of paper by following the steps below.

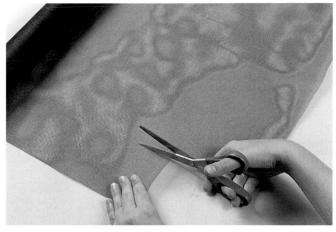

2 Use the scissors to cut a piece of fiberglass screening that is large enough to wrap around the side and the top of the frame.

3 Tuck the fiberglass edges underneath the hardware cloth, and staple it to the front of the wooden frame. Use the hammer to flatten any staples that need it.

1 Seal the wooden frames with the wood sealant, and allow them to dry thoroughly.

4 Cover the edges of the frame with the duct tape. Avoid covering any parts of the exposed screen.

HOW TO PULL A SHEET OF PAPER

The following steps and photos will show you how easy it is to pull a sheet of paper using a mould and deckle.

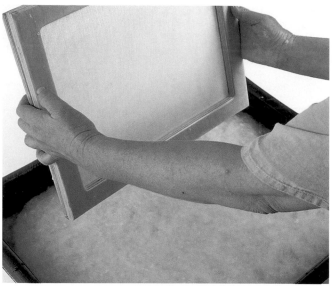

2 Position the mould with the screen side up. Place the deckle on top of it. Hold the mould and deckle together by the wooden edges. (Avoid placing your fingers over the screen, or you'll create holes in your paper.)

1 Fill your vat half full of water, and add several handfuls of pulp. Stir it well to evenly distribute the pulp in the water.

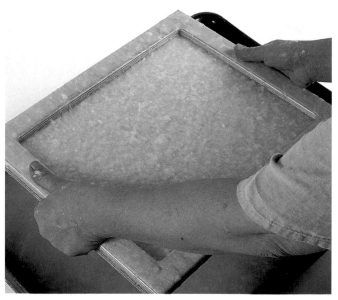

3 Beginning at the far side of the vat, use a scooping motion to dip the mould and deckle and gather pulp. Bring it up out of the vat in a horizontal position. (If you tilt it, some of the pulp will slide down the screen and make an uneven sheet of paper.) Hold the screen in a horizontal position, and allow the water to drain through for a few moments, letting the pulp settle on the screen. If the water is draining too slowly, rub a sponge across the bottom of the screen to absorb the water.

4 Place a damp felt or cloth on your worktable. Place the mould and deckle beside it. Carefully remove the deckle, and set it aside.

5 Rest the mould with the paper faceup along the bottom edge of the felt. Then slowly flip it face-down onto the felt.

6 Sponge any excess water from the back of the screen to release the sheet from the mould.

7 Hold one edge of the mould down and lift the other edge of the mould up, as if you're opening out the cover of a book. Check to be sure that the paper is releasing well. If it isn't releasing, sponge the back of the screen again. If it's a very thin sheet, and it isn't releasing, you might need to pour water back on the mould and press it down to make good contact with the felt. Then, using a rocking motion, remove the mould from the paper. Once the paper is released, lay another felt on top and press well with the sponge to release more water. Carefully remove the paper from the felt, and gently brush it onto a drying surface such as a sheet of countertop material, glass, or mirror. Allow the paper to dry completely before peeling it off the surface.

COMMONLY ASKED QUESTIONS

THE FOLLOWING QUESTIONS are often asked by students who are beginning to learn about papermaking.

HOW DO YOU KNOW HOW MUCH WATER AND PULP TO MIX TOGETHER?

Begin by filling your vat about half full of water. This amount of water leaves enough room for dipping the mould and allows pulp fibers to float freely without crowding them. Begin by adding a few handfuls of pulp to the water. Stir it thoroughly to distribute it evenly before pulling a sheet of paper. To help you decide whether the sheet you've pulled is

Adding inclusions to a vat of water

the correct thickness, keep in mind that the pulp will be pressed and flattened after you couch it. As you gain experience, you'll develop an eye for how much pulp you need to make certain kinds of paper.

You'll be able to pull a thicker sheet by using the deckle with the mould, since the deckle helps to contain the pulp on the screen. If the sheet looks too thin, add another handful or more of pulp to the mould. When using up the thin remains in a vat, use a deckle to contain the pulp and laminate several thin sheets together to make thicker ones.

It's easier to make thin sheets if you remove the deckle and use the screen alone. If the sheet still looks too thick, remove some of the pulp from the vat so that you don't bring up a lot of pulp on your mould. To accomplish this, scoop it out with the wrong side of the mould before pulling another sheet.

HOW DO YOU CARE FOR YOUR MOULD AND DECKLE?

Clean it with cool water (preferably dispensed through a hose with a nozzle) after you've finished a session of papermaking. Dry and store the mould and deckle on its side. Avoid storing it in a hot place with direct sunlight.

HOW LONG WILL THE PULP STAY FRESH?

Paper pulps are vegetable matter, so they have a fairly short life before they begin to break down. Nevertheless, there are some things that can be done to prolong their usability.

Cotton and recycled pulps will keep for a week or two if they're stored in a cool room. Abaca doesn't last as long and, like all pulps, has a very unpleasant odor when it starts to break down. Continue to use pulps for as long as you can stand working with them—they won't smell after they've dried. You can extend the life of pulps by storing

Refrigerate cooked plants and prepared pulp in plastic bags to increase their life

them in a closed container or plastic bag in the refrigerator. If you're using pulp that contains no sizing, you can extend the life of it in a couple of ways:

1. Use a colander to drain the water from the pulp, and store it in a plastic bag or container in the freezer. Reprocess the pulp after it thaws out.

2. Use a colander to drain the water out of the pulp before placing it on a screen that allows the water to drip out of it while it dries (do this outside or over a vat). You can also squeeze the balls of pulp by hand to remove the water before allowing them to dry. Store the dried pulp in a paper bag. When you're ready to use it, soak it in water to rehydrate it before processing it again to loosen the fibers. Be sure to label all your stored pulps!

HOW DO YOU MAINTAIN AND CARE FOR YOUR FELTS?

All felts—whether cotton sheeting, muslin, nonwoven interfacing, or wool felts—need to be kept clean. Brush them off to remove bits of pulp from the surface before you reuse or store them. Hang them on an indoor or outdoor clothesline to dry, or dry them in a clothes dryer (if using cotton fabric or nonwoven interfacing) to prevent them from mildewing after they're stored.

ADDING INCLUSIONS TO HANDMADE PAPER

INCLUSIONS, OR MATERIALS ADDED to the pulp, can change the color and texture of your papers. Dried flowers, leaves, seeds, rice, sand, potting soil, threads, and other materials can be added directly to the vat of pulp.

Experiment with different inclusions and do some trial runs on how much of a particular material or materials to add to achieve the results you want. You can use either of the following simple techniques to add inclusions to handmade paper:

1. Place inclusions in water until they are thoroughly saturated. Prepare to pull a sheet of paper by adding pulp to a vat of water, and stir the vat of pulp as you normally would. Add the hydrated inclusions to the surface of the vat without any further stirring. Use a mould and deckle to pull a sheet of paper. The inclusions will be scattered through the pulp of the final piece of paper, and, for this reason, some of them will be covered up after the paper dries.

2. Prepare to pull a sheet of paper by adding pulp to a vat of water. Fill a second vat with water, and add a handful of the same pulp and inclusions. Pull a sheet of paper from the first vat, and couch it onto a felt. Stir

Flowers, leaves, and potting soil leave random patterns in paper.

the inclusion vat well, and pull a thin sheet from this vat. Couch it on top of the first sheet so that the surface of your paper has a laminated layer of inclusions. With this method, you can fill several vats with different inclusions and create different papers with the same vat of pulp.

Shredded papers, plants, seeds, and potting soil can be used as inclusions.

TECHNIQUES

YOU CAN USE SEVERAL techniques to alter the surface of handmade paper. Use them in combination or alone.

Masking

By masking the screen of your mould with masking tape, duct tape, or contact paper, you can accomplish several things. First, you can apply tape to the topside of the mould's screen, and mask off areas in order to create shaped papers. For instance, if you simply run a length of tape down the middle of a rectangular mould and divide it in half, you'll create two separate pieces of paper when you pull the paper. You can mask off shapes that will translate into envelopes by using an unfolded commercial envelope as your template.

To create a grid pattern of windows on your paper, pull a sheet of plain paper, and couch it on a felt. Then mask off the mould with tape to create a window pattern. Pull another sheet that is thinner than the first, and laminate this one on top of it.

After pulling and couching a sheet of plain paper, mask off your screen with tape in a design of your choice. Pull a second, thinner sheet and laminate it on top of the first to create a surface pattern.

Embossing

You can add both texture and pattern to paper by impressing materials on its surface. Any flexible, unbreakable, flat materials can be used to emboss paper—twigs, bamboo skewers, bath mats with patterns, string, cords, corrugated cardboard, buttons, or coins.

To emboss paper, place the materials on the surface of couched, wet paper. Arrange the embossing material or materials facedown on the paper, keeping in mind the design that they will leave after embossing.

If the material that you're using is large enough to cover the entire sheet of paper, sponge directly and evenly on top of it. If you're embossing with several small items, cover them with a sheet of fiberglass screening before pressing them with a sponge.

Handmade paper can be embossed with any flat, water-repellant material.

Stitching

Hand or machine stitching on handmade paper offers wonderful opportunities for texturing, embellishment, and layering. Your papers should be thoroughly dry before any piercing or stitching is attempted.

To hand stitch paper, lay the paper over a piece of flat scrap board. Use an awl to pierce the paper just enough for the sewing needle to pass through it easily. You can stitch with a sharp needle without piercing the paper first, but the paper has a tendency to tear.

To machine stitch paper, cut a piece of wax paper or nonwoven interfacing that is slightly smaller than the paper, and place it underneath it. This will assist you in easing the papers over the bed of the sewing machine and will serve as a support for the sewn paper. Use a series of freeform stitches of various sizes and widths, or create patterns with straight stitches. You can also create an interesting surface on the paper by piercing it with an unthreaded needle.

Embellishing

Stitch small trinkets and found objects such as beads, buttons, shells, small rocks, or twigs to the surface of your paper to embellish it. To make the stitching easier, position the object you wish to stitch on the paper. With an awl, pierce the paper where the stitches will go before you attach it. If you do this first, sewing will go more quickly, and you won't have to force a needle through the paper and risk tearing it.

Embellish your handmade paper piece by sewing on beads and other found objects.

Hand or machine stitching can be added to paper to create patterns and designs.

Waxing

Waxing lends the surface of handmade paper a translucent, beautiful look. The choice of waxing the surface of your paper has several advantages. The application of wax will deepen the colors and create a wet appearance. If

Paint on a coat of melted parrafin or beeswax to enhance the finish and highlight the inclusions on your paper.

you've added inclusions to your papers that disappeared after the paper dried, wax will bring them out.

Always work in a well-ventilated area with an exhaust fan when you're waxing. Paraffin and beeswax are flammable, and the fumes that result from melting them aren't healthy to breathe. Use an electric skillet with a temperature gauge set at 300°F (149°C). At this temperature, paraffin or wax will melt without smoking. If it does smoke, it's too hot. Make sure to cover the skillet when you aren't using it.

To wax paper, gather the following materials and tools: paraffin (from the grocery store) or beeswax (from a candle or textile arts supplier), an electric skillet and clothes iron that are designated for waxing, a pair of tongs, a natural bristle paintbrush, newspapers, and newsprint.

Cover your work surface with a thick blanket of newspapers followed by sheets of unprinted newsprint. Melt the paraffin or beeswax thoroughly. If your paper fits easily into the skillet, carefully dip it in and out using the tongs. (Tongs also come in handy if you drop anything into the wax by accident.) If the paper doesn't fit, position it on the newsprint, and use the brush to paint on a layer of paraffin or wax. Allow it to dry for a couple of minutes before waxing the other side.

Sandwich your paper between two sheets of newsprint, and iron out the wax. (You'll need to change the newsprint several times before you're finished.) Again, be careful not to breathe the fumes that will result from this process.

GALLERY

Top left: **EMILY TUTTLE,** *House Cat,* Collagraph print on handmade paper, 20 x 16 inches (50.8 x 40.6 cm)

Top right: **ARLYN ENDE**, *Could Dinsky?,* Mixed media collage with handmade paper, 30 x 22 inches (76.2 x 55.9 cm)

Right: **GAIL LOOPER**, *My Rock n' Roll Honey,* Mixed media collage on watercolor paper with handmade paper, concert and raffle ticket stubs, thread, waxed linen, and machine stitching, 7 x 17 inches (17.8 x 43.2 cm)

Top left: **ARLYN ENDE**, *A Shrine of Language*, Mixed media collage with handmade paper, acrylic paint, and gold leaf, 22 x 18 inches (55.9 x 45.7 cm), Photo by Jim Ann Howard.

Top right: **ANN HARTLEY**, *Road to India 1*, Mixed media collage with handmade paper, 19 x 17 inches (48.3 x 43.2 cm)

Left: **SANDY WEBSTER**, *References*, Assemblage with handmade paper (that contains inclusions such as sawdust, tobaccos, produce, denim, and want ads), found materials, and other media, 12 x 18 x 9 inches (30.5 x 45.7 x 22.9 cm)

Top center: **CLAUDIA LEE**,
Table Screen, Handmade paper,
15 x 34 inches (38.1 x 86.3 cm)

Above left: **C. CARLENE
TAYLOR**, *Journal*, Handbound
book on handmade paper with
Coptic binding and copper,
5½ x 8 inches (14 x 20.3 cm)

Right: **EMILY TUTTLE**, *Stream
of Consciousness*, Collagraph
print on handmade paper,
20 x 8 inches (50.8 x 20.3 cm)

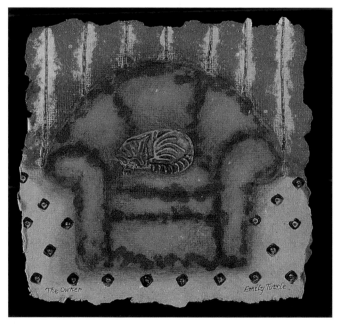

Upper left: **EMILY WILSON**, Handmade box covered with handmade and commercial papers, 5½ x 6 x 1¾ inches (14 x 15.2 x 4.4 cm)

Upper right: **CLAUDIA LEE**, *Cactii*, Assembled handmade paper pieces with painted embellishment, 25½ x 23 x 1 inch (64.8 x 58.4 x 2.5 cm)

Lower left: **EMILY TUTTLE**, *The Owner*, Collagraph on handmade paper, 8½ x 8½ inches (21.6 x 21.6 cm)

Lower right: **CLAUDIA LEE**, Papier-mâché bowl assembled from scraps of handmade paper, embellished with painted designs, 8¾ x 2½ inches (22 x 6.4 cm)

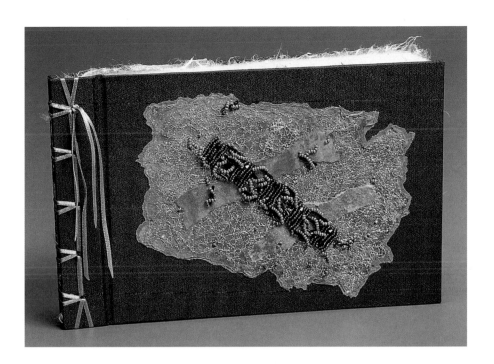

Top center: **GAIL LOOPER**, *Silvery Dreams*, Japanese bound book with handmade paper panel, embellished with silver metallic thread, beaded loomwork, and ribbon, 5 x 10½ inches (12.7 x 26.7 cm)

Lower left: **C. CARLENE TAYLOR**, *Present Edition*, Handbound book with handmade paper cover, 4 x 6 inches (10.2 x 15. 2 cm)

Lower right: **CLAUDIA LEE**, *Books Within Books*, Handcolored papers, screen printing, collage, 8 x 18½ inches (20.3 x 47 cm)

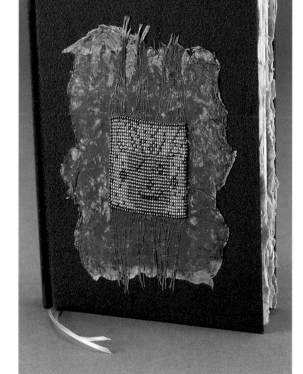

Top center: **CLAUDIA LEE**, *Out of Time*, Handmade paper, stitching, collage, 15½ x 11 x 2½ inches (39.4 x 27.9 x 6.4 cm)

Lower left: **EMILY TUTTLE**, *Equal Opportunity*, Collagraph print on handmade paper, 16 x 20 inches (40.6 x 50.8 cm)

Lower right: **GAIL LOOPER**, *Joni's Book*, Western bound book with handmade paper panel on front cover embellished with beaded loomwork, 6½ x 8 inches (16.5 x 20.3 cm)

Straw Paper

DESIGN: **CLAUDIA K. LEE**

By adding straw pulp to paper, you'll create a textured, lively surface.

YOU WILL NEED

Hanging or baby scale for weighing plant materials

Dry plant materials, fresh or dried (see page 11 for suggestions)

Two large plastic buckets

Water source

Large stainless steel pot

Hot plate or other heating source

Rubber gloves

Plastic goggles or safety glasses

Gram and ounce scale for measuring soda ash

Soda ash (an alkali)

Wooden stirring rod or length of PVC pipe

Large colander (available at restaurant supply stores)

Pot holders

Hand tool for blending the pulp (such as a paddle, wooden mallet, or baseball bat) or electric blender

Hard, water-tolerant surface such as a sheet of kitchen counter material, plywood, or a wooden cutting board (if beating the pulp by hand)

Clear quart (.95 L) jar with lid

Blender

Kitchen scissors or pruning shears (if using a blender to blend the pulp)

Plastic vat or tub

Mould and deckle

Fabric to use as a felt such as a sheet of wool papermaking felt (see papermaking supplier on page 79), an old blanket, cotton sheeting, non-woven interfacing, or muslin

Sponge

3-inch (7.6 cm) soft-bristled paintbrush

Drying surface such as sheet of light diffuser grid (plastic), countertop material, glass, or mirror

SAFETY TIPS FOR PROCESSING PLANT MATERIALS:

- Don't do this in your kitchen because some of the plants and chemicals are toxic if ingested. Instead, set up a hot plate in an area where you don't prepare or eat food.
- Don't use tools or pots that you also use in the kitchen.
- Don't use aluminum pots or measuring tools that can react with alkalis.
- Work in a well-ventilated area.
- Add the alkali (soda ash) to the pot of water before it boils to avoid splashing (see below).
- Add the fibers to the pot after the soda ash has been stirred in and it has dissolved. (It will dissolve quickly.)

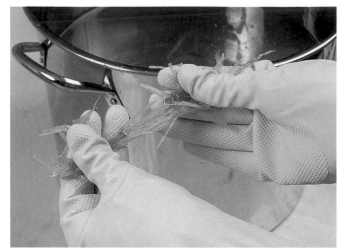

2 Check the plant pulp for doneness every half hour or so by removing some of the fiber and pulling it apart against the grain. When it is done, it should pull apart easily.

1 Use the hanging or baby scale to weigh out enough dry plant materials to fill half of your bucket. Record how much the materials weigh. Place them in one of the buckets, and cover them with water. Soak them for several hours or overnight. Pour several gallons of water, or enough to cover the plant materials, into the stainless steel pot. Place it on the hot plate or other heating source, and heat the water to just below a boil. Put on rubber gloves. Use the gram and ounce scale to weigh out an amount of soda ash that is equal to approximately 20 percent of the dry weight of the fibers that you recorded. (For one pound [.45 kg] of dry plant fibers, use 3½ ounces [100 g] of soda ash.) Add the soda ash to the pot, and stir it in gently with a stirring rod. (It only takes a few minutes for the soda ash to dissolve.) Add the soaked plant material to the pot and stir. Allow the water to come to a boil. Reduce the heat, and allow the contents to simmer.

3 Place the colander on top of an empty bucket. Grip the stainless steel pot with pot holders, and pour the water and pulp into the colander. Rinse the pulp thoroughly with cool water from a hose or other source.

4 To beat the pulp by hand, use the paddle, wooden mallet, baseball bat, or other tool. Pile the cooked fibers on top of the hard, water-tolerant surface. Beat them from one direction with the tool in order to begin flattening them. After flattening the whole pile, rotate them 90°, and repeat this process. If the fibers become slightly dry, pour some water on them, and continue beating them. When you think that the fibers are beaten to a pulp, test them by dropping a spoonful of them into the clear quart jar filled with water. Put the lid on, and shake well. If the fibers disperse evenly, without clumping, your pulp is ready. If not, beat them some more until they disperse in the water.

6 Add the pulp to a vat or tub filled with water. Immerse the mould and deckle. Pull a sheet of paper by lifting the mould and deckle in a horizontal position from the tub.

7 Couch the paper onto the felt, and press with the sponge.

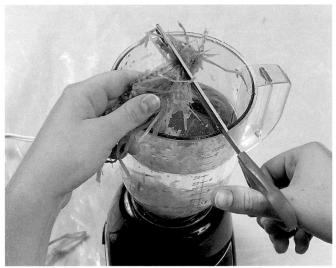

5 You can also process the fibers into pulp with a blender. To do this, fill the blender about three-quarters full of water. Use the kitchen scissors or pruning shears to cut the fibers into smaller pieces that can be handled by the blender. Place them in the blender without overloading it. Blend the fibers in spurts. Remove some of the fibers or add more water if the blender gets clogged.

8 Carefully peel the finished sheets off the felt, and gently brush them onto a drying surface such as a sheet of glass or countertop material. Allow them to dry in a well-ventilated area overnight, or until completely dry, before peeling them off the surface.

Spirit Money Paper

DESIGN: **CLAUDIA K. LEE**

Make gorgeous gold and mauve paper by recycling Asian spirit money. You can transform an embroidery hoop into a mould to make elegant circular papers, or use a traditional mould and deckle for this project.

YOU WILL NEED
Spirit money or joss papers (available in Asian groceries)
Water source
Blender
Small plastic tub or vat just large enough to accommodate embroidery hoop or mould and deckle
Embroidery hoop fitted with sheet of fiberglass screening to serve as a mould or a traditional mould and deckle
Fabric to use as a felt such as a sheet of wool papermaking felt (see papermaking supplier on page 79), an old wool blanket, cotton sheeting, non-woven interfacing, or muslin
Sponge
Drying surface such as sheet of light diffuser grid (plastic), countertop material, glass, or mirror
3-inch-wide (7.6 cm) soft-bristle brush

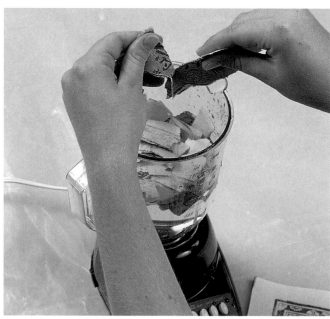

1 Fill the blender one-half to three-quarters full of water, and tear the spirit money papers directly into the blender. Because spirit money or joss papers break down easily in water, you don't need to presoak them.

2 Engage the blender in spurts. Pay attention to any indications that the blender is stressed. If you smell the blender overheating, turn it off, and remove some of the fibers before adding more water. Blend as much or as little as you wish, to create the consistency and look that you want.

3 Fill the tub or vat three-quarters full of water, and add the pulp from the blender. To begin pulling a sheet of paper, dip the embroidery hoop or mould and deckle into the water and pulp. (The hoop will create round sheets of paper.)

4 Pull a sheet of paper, and allow the excess water to drain into the vat. Couch and sponge the paper onto the felt, and remove the hoop.

5 These sheets are rather delicate and can be difficult to handle. For this reason, turn the felt, with the wet paper still attached, facedown onto your drying surface. Press down on the back of the felt with the sponge to make good contact, and carefully peel up the felt.

6 Use the brush to gently spread the sheet onto the drying surface, paying special attention to the edges. Allow the paper to dry thoroughly before peeling it up.

Sewing Pattern Paper

DESIGN: **CLAUDIA K. LEE**

Who would've ever dreamed that you could make beautiful paper from old sewing patterns?
They recycle into luscious, butterscotch-colored papers with flecks of black.

YOU WILL NEED

Tissue paper sewing patterns

Scissors or paper cutter

3 large plastic buckets

Water source

Blender

Colander

Plastic vat or tub

Mould and deckle

Fabric to use as a felt such as a sheet of wool papermaking felt (see paper-making supplier on page 79), an old wool blanket, cotton sheeting, non-woven interfacing, or muslin

Sponge

Drying surface such as sheet of light diffuser grid (plastic), countertop material, glass, or mirror

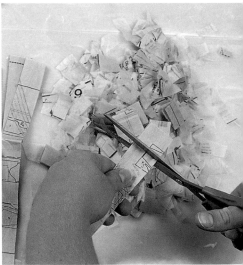

1 Use scissors or a paper cutter to cut the patterns into small pieces that are approximately an inch (2.5 cm) square.

2 Soak the cut papers in a
  bucket of water for a few
minutes or until they are thoroughly saturated.

3 Fill the blender half full of
  water. Add the soaked paper in
small handfuls, and place the lid
on securely. Turn the blender on,
and begin blending with intermittent spurts before turning it on
high. During this process, listen as
well as look at what's happening. If

there's plenty of movement, you
may want to add more fibers. If
there isn't much movement, and
things seem to be clogged, stop the
machine, and remove some of the
fibers before adding more water. If
you smell the motor overheating,
stop the blender immediately, and
allow it to cool down. (Don't forget
that blenders were not made for
processing paper!) Remove some of
the fibers, and add more water
before blending again.

4 Keep checking the pulp, and
  when it is blended until it is
the consistency of oatmeal, turn
off the machine, and dump the
contents of the blender into a
colander to drain over another
bucket. Empty the drained pulp
into the other bucket. Continue
this process until you have enough
pulp to make the amount of paper
that you need. If you run short, it's
easy to make more and continue
working.

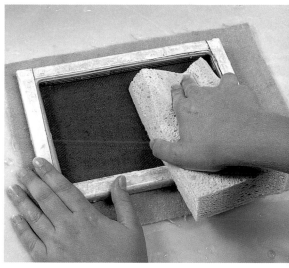

5 Add the pulp to the vat or
  tub of water, and stir it well.
Pull a sheet of paper with a
mould and deckle. Remove the
deckle and couch the paper onto
a felt. Sponge away the excess
moisture from the back of the
mould to release the paper.

6 Remove the mould. With the
  felt intact, flip the sheet of
paper over onto a drying surface
such as a sheet of glass. Sponge
the back of the felt again. Peel
away the felt. Allow the paper to
dry overnight.

Marbled Paper

DESIGN: **CLAUDIA K. LEE**

Marbling makes a great way to use up small amounts of colored pulps to create colorful papers.

YOU WILL NEED

Several drained pulps with sizing added (see page 12) of different colors (use 2 to 4 cups [.48 to .96 L] of each color to make 10 to 15 sheets of paper)

Plastic vat or tub

Water source

4 to 6 cups (.96 to 1.4 L) of plain, uncolored pulp

Colander

Several buckets to hold various colors of pulp

Clear glass quart (.95 L) container or blender

Set of measuring spoons

PNS formation aid (see page 79 for supplier)

Wire whisk or other tool for stirring

Mould and deckle

Fabric to use as a felt such as a sheet of wool papermaking felt (see papermaking supplier on page 79), an old wool blanket, cotton sheeting, non-woven interfacing, or muslin

Sponge

Drying surface such as sheet of light diffuser grid (plastic), countertop material, glass, or mirror

1 Place each drained colored pulp in a separate bucket. Fill

the plastic tub with water, and add the plain pulp. Set aside the plain pulp. If you're using the clear quart (.95 L) container, run cool water in a slow, steady stream to fill the container. Slowly add ½ teaspoon (2.5 ml) of PNS powder with one hand while stirring with the other. If you're using a blender to mix the PNS and water, whir them together until thoroughly mixed.

2 Add a small amount of the mixture (around a tablespoon [15 mL] to 2 cups [.48 L] of pulp) to each of the colored pulps.

3 Mix the pulp and PNS together with your hands until the mixture feels gooey. Repeat this process with as many colors of pulp as you wish to add to use for your marbled paper.

4 Add handfuls of color to the vat or tub of plain pulp. The colored pulps will flocculate or cluster rather than blending with the pulp, creating a marbled look.

5 Use the mould and deckle to pull a sheet of paper from the marbled mixture.

6 Couch, sponge, and press the paper onto the drying surface.

Embossed Paper

DESIGN: **CLAUDIA K. LEE**

Use scraps of everyday materials to create texture and pattern on your paper.

YOU WILL NEED

Items for embossing that can stand up to water (if you plan to use a press, these items should be nonbreakable): chair caning, bath mats, buttons, patterned polystyrene foam containers, crocheted doilies, mesh bags, shaped wire, foamcore, insulation board cut into shapes, corrugated cardboard, sticks, textured place mats, textured fabric, textured wallpaper samples, grass, wrinkled paper, rope, cords, etc.

Scissors (optional)

Pulp made from cotton linter (see pages 11 and 12)

Plastic vat or tub

Mould and deckle

Fabric to use as a felt such as a sheet of wool papermaking felt (see papermaking supplier on page 79), an old wool blanket, cotton sheeting, non-woven interfacing, or muslin

Sponge

Sheet of fiberglass screening (optional)

Small bookbinding press and felts (optional)

Drying surface such as sheet of light diffuser grid (plastic), countertop material, glass, or mirror

1 Collect the materials that you plan to use for embossing. Use them as they are, or cut them into interesting shapes. Select items that can be easily hand pressed or put through a press without breaking.

2 Fill the vat with water, and add the cotton linter pulp. Form some thick sheets of paper with the mould and deckle. (Or laminate and couch thinner sheets together to form thick sheets.) Arrange the embossing materials facedown on the wet sheets of paper, keeping in mind the design that they will leave after embossing. If the material that you're using is large enough to cover the entire sheet of paper, sponge directly and evenly on top of it.

3 Peel the materials back from the surface to reveal the embossed impressions on the paper.

4 If you're embossing with several small items, cover them with a sheet of fiberglass screening before pressing them with a sponge.

NOTE: If you have a small press that you plan to use, lay a felt or other pressing cloth on top of the embossing materials and paper before running it through the press. If you want to emboss both sides of the paper at the same time, lay the embossing materials on the felt before couching the paper on top, and finish with more embossing materials before pressing. If you plan to press a stack of several embossed pieces at once, separate the papers with several felts to prevent them from embossing into the sheets above and below them.

5 Carefully remove the screening and the embossing materials from the paper. Transfer the papers gently to a drying surface and brush them into place.

Handmade Papers with Added Color

DESIGN: **CLAUDIA K. LEE**

You can spray colored paints onto any kind of handmade paper to enhance and alter it. We used paper made from recycled pulps, but any cream, white, or light gray paper works well for showing off colors.

Empty spray bottles (a garden-supply store is a good place to find these)

Water source

Water-based coloring mediums such as acrylic paints, poster paints, or pigments for papermaking

Scrap paper

Sheet of plastic large enough to cover work surface (a plastic dropcloth works well)

Large sheets of newsprint or newspaper

Freshly pressed, damp sheets of lightly colored or uncolored handmade paper

Rubber gloves

Apron

Items for masking the paper such as string, yarns, cutout paper shapes, hardware, scraps of cloth, leaves, etc.

Piece of lightweight cotton fabric such as an old bedsheet or cloth diaper (optional)

Clothes iron (optional)

1 Fill empty spray bottles with water and add coloring mediums of your choice. Begin with about a tablespoon (15 mL) of color. The more color you add, the more intense it will be. Test out each color that you mix on a sheet of scrap paper to see if you like the strength of the color.

2 Spread the plastic over your work surface followed by a couple of layers of newsprint or newspaper. Place a sheet of freshly pressed handmade paper on your paper-protected work surface. Put on the rubber gloves and apron to protect your hands and clothing.

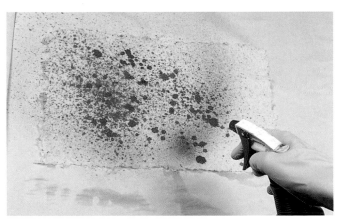

3 Randomly spray the surface of your paper with one color followed by another. (The colors will mix as they overlap. You might want to test out your choices on a sheet of scrap paper first.)

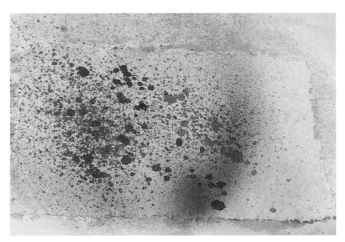

4 Add other colors on top of your first layer of colors, creating the look of a watercolor. Notice how the addition of other colors will tint and change the appearance of the colors underneath.

5 Lay items of your choice on top of the surface to mask it. Spray the surface with more color.

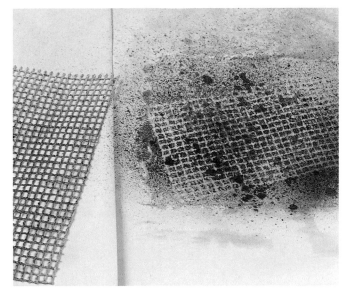

6 Remove the masking items to reveal patterns underneath. If you like, you can keep adding layers of masked shapes to your design. Repeat this process of spraying and decorating the surface on the other side of the paper, if you wish. When you're finished, you can dry the paper in several ways. If you want, you can iron it to dry it out. To do this, place a sheet of light-weight cotton fabric on top of the surface of your paper, set your iron on a cotton setting, and press it to remove the moisture. If you allow the paper to air dry and it warps, spray the surface lightly with water from a spray bottle to relax it before ironing it.

Paper Ornaments

DESIGN: **CLAUDIA K. LEE**

These playful ornaments make a fun, easy project for children as well as adults.

¼-inch (6 mm) ribbons, each 21 inches (53.3 cm) long

5-inch (12.7 cm) weaving needle, or any long, large-eyed needle

3-inch (7.6 cm) polystyrene foam balls

Two cardboard boxes of the same height, each at least 18 inches (45.7 cm) high

36-inch-long (91.5 cm) wooden dowel

Masking tape

Several sheets of freshly pressed papers, still wet

1-inch-wide (2.5 cm) paintbrush

Methylcellulose (a mild glue purchased at art-supply stores or from papermaking suppliers) or wallpaper paste (purchased at home-supply stores)

Fabric to use as a felt such as a sheet of wool papermaking felt (see papermaking supplier on page 79), an old wool blanket, cotton sheeting, non-woven interfacing, or muslin

Star-shaped cookie cutter without a bottom or top

Wet, colored pulp of your choice

Pencil or craft stick

Piece of fiberglass screen cut into a square larger than the star cutter

Sponge

Clothes iron (optional)

Flat scrap of wood

Ice pick or awl

Sponge brushes

Assorted acrylic paints

TIP: You can make more sheets than you need, and leave them between wet felts overnight so they are usable later.

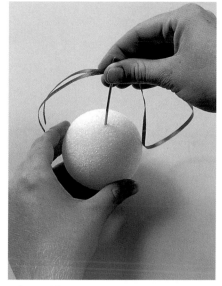

1 Fold one of the ribbons in half with a loop on one end and two loose ends on the other. Thread the ends through the needle. Push the needle and ribbon through the center of one of the balls and out the other side.

2 Tie an overhand knot in the ends of the ribbon. Gently tug it until it pulls into the ball and leaves the bottom of the ball smooth. Your ball will now have a looped hanger attached to it. In order to apply papers to the ball, you'll need to suspend it by its

hanger so that nothing is touching it. One simple solution is to place two cardboard boxes on a table, leaving space between them for suspending the ball. String the loop of the ball's hanger onto a wooden dowel so that it hangs freely, and place the dowel on top of the boxes so that it forms a bridge between them. To steady the dowel, tape the ends of it into place with masking tape. If you want to work on several ornaments at once, you can string several balls with ribbons and suspend them on the dowel about 6 inches (15.2 cm) apart.

3 Tear the moist, handmade papers into smaller strips and pieces. Apply the papers in layers to the ball, and brush them into place with lots of glue. Continue gluing the paper, overlapping the edges as you go. Cover the whole ball with the paper.

4 Lay the cookie cutter on the felt. Scoop out a small handful of pulp and push it into the center of the cutter. Use the rubber end of a pencil or a craft stick to press the pulp into a thin layer. Gently lift the cookie cutter.

6 Place the paper star on the flat scrap of wood, and punch a hole in the center of it with the ice pick or awl.

7 Slide the star down over the ribbon, and brush it into place with glue and the brush. This can be done when the ball is still wet.

5 Lay the piece of screen over the star, and use the sponge to press the water from it. Use a warm clothes iron to dry the star, or allow it to air dry on the felt or on a drying rack or countertop.

8 When the ball is thoroughly dry, use sponge brushes loaded with acrylic paints to decorate the surface. Allow the ball to hang and dry overnight.

Handmade Paper Envelopes

DESIGN: **CLAUDIA K. LEE**

Decorate the outside of a handmade paper envelope with a handmade paper button or other unusual closure.

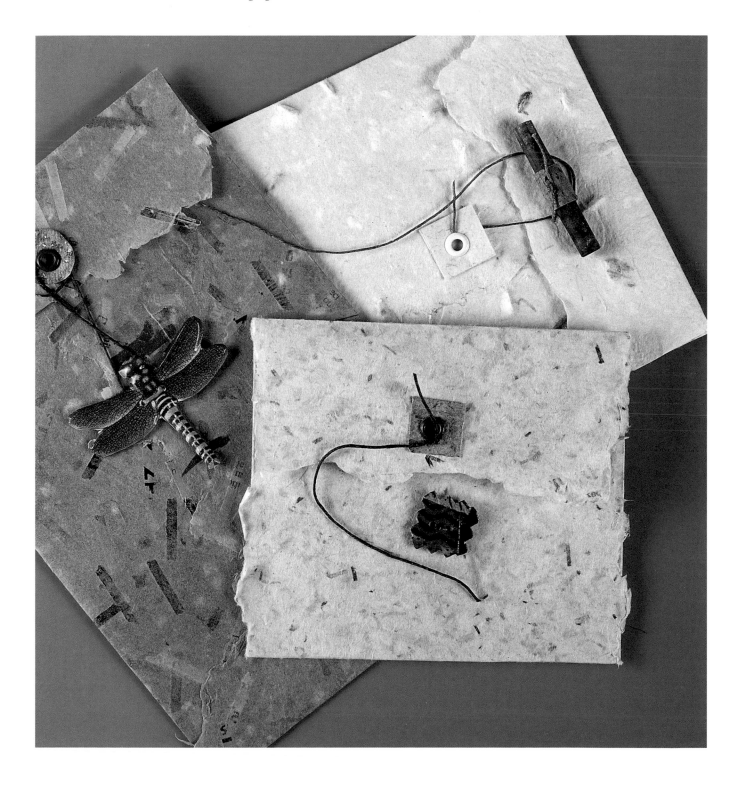

Commercial envelope in size and shape of your choice

Mould large enough to accommodate the envelope when it is opened out

Mat board, thin cardboard, or poster board

Scissors

Craft knife with sharp blade

Sheet of insulation board (to be used for the shaped deckle)

Marking pen

Plastic vat or tub

Pulp of your choice

Fabric to use as a felt such as a sheet of wool papermaking felt (see paper-making supplier on page 79), an old wool blanket, cotton sheeting, non-woven interfacing, or muslin

Drying surface such as sheet of light diffuser grid (plastic), countertop material, glass, or mirror

Metal ruler

Bone folder (a bookmaking tool found at an art-supply store) or butter knife

Needle and thread

Paper button, regular button, or other object to use as a closure (See the project on pages 48 to 49 for instructions on how to make paper buttons.)

Scissors

Double-sided tape

Small pieces of decorative paper

White craft glue

Hole punch

Eyelet kit with flange

Flat scrap of wood for hammering surface

Hammer

Length of waxed linen or thin cord

1 Carefully open out the envelope by pulling apart the glued edges. (This pattern will be used to create a shaped deckle for your handmade envelope.) Choose a mould that is large enough to accommodate the envelope pattern. Trace the opened envelope onto a piece of mat board, thin cardboard, or poster board. Cut out the shape with scissors. With the craft knife, cut a piece of the insulation board the same dimension as your mould. Trace the envelope template onto the insulation board with a marking pen, leaving a margin of at least an inch (2.5 cm). Cut out the envelope shape carefully with a sharp craft knife. Place the shaped deckle on top of your mould. Fill the vat with water and a pulp of your choice. Pull as many envelopes as you like. (Don't make the sheets too thick since you will be folding them.) Couch, press, and dry them as you would any sheet of paper.

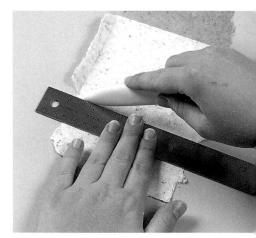

2 Place the metal ruler along one of the fold lines of the envelope, and score the line with the bone folder. Repeat this process for the other three fold lines. Place the bone folder along each fold line, and crease the flaps. Fold in the short side edges followed by the bottom flap and the top flap.

3 Open out the bottom flap. Use a needle and thread to stitch the paper button or other closure in the center of the bottom flap just underneath the edge of the top flap (see the photo of the finished piece for guidance).

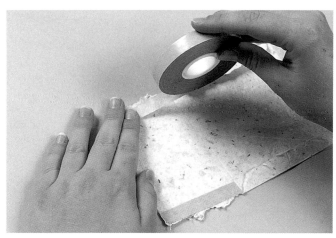

4 Place a strip of double-sided tape along each of the inside edges of the bottom flap of the envelope. Tape the bottom flap into place with the button on the outside.

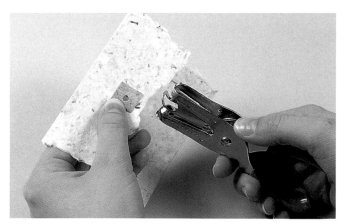

5 Cut out a small square of decorative paper. Glue it down in the center of the top flap above the button that you sewed to the bottom flap. With the hole punch, punch a hole through the middle of the paper and flap.

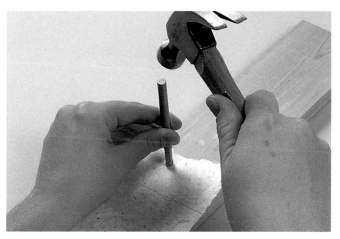

6 Place an eyelet in the punched hole on the front of the envelope. Open the top flap out, and place it, right side down, on the flat scrap of wood. Use the flange that comes with the eyelet kit and a hammer to tap the eyelet into place on the wrong side of the flap.

7 Cut a length of waxed linen thread or cord at least 5 inches (12.7 cm) long. You can trim it later. Hold the short end of it next to the eyelet with the thumb of one hand. Pull the long end around and underneath the eyelet with your other hand. The eyelet will hold it in place. Close the envelope and wrap the long end of the cord around the closure. Trim the cord to an appropriate length.

TIP: You can make a simple glue known as "re-lick" glue for sealing envelopes by mixing two parts of white craft glue with one part of vinegar. This glue is safe to lick! After you've mixed the ingredients, brush the glue on with a small brush, and allow it to dry completely before applying a second coat.

Stitched Paper Collage

DESIGN: **CLAUDIA K. LEE**

Making a collage is a simple, accessible way to make an engaging piece of artwork.

YOU WILL NEED

Scraps of handmade and decorative papers

Scissors

Square-shaped sheet of sturdy hand-made paper in a neutral color on which to mount the collage (foundation sheet)

White craft glue

Larger sheet of square-shaped, off-white, handmade paper to serve as a background for the finished collage

Sewing machine

Spray matte finish or gloss medium

Small paintbrush (optional)

Sheet of wax paper

Small stack of books

An assortment of beads, buttons, and other trinkets

Needle and thread (optional)

1 Cut scraps of papers into interesting shapes. Begin by placing the paper shapes in different configurations on your foundation sheet. Don't glue them down yet. Play around with them until you have a design that appeals to you.

2 When you're satisfied with your basic design, glue the pieces into place with white craft glue. Allow the glue to dry.

3 Place and center the collage on top of the background paper. Use the sewing machine to sew a seam around the edge of the collage to hold it in place on the paper.

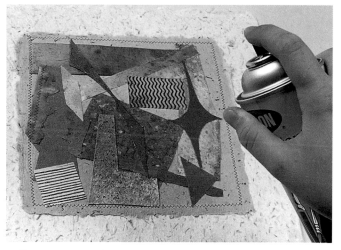

4 Coat your finished piece with spray matte finish, or paint on a coat of gloss medium. Allow the finish or medium to dry thoroughly. Place a sheet of wax paper on top of the collage, stack books on top of it, and allow it to sit overnight.

5 Glue or sew on buttons, beads, or other trinkets to create more surface texture.

Collaged Paper Buttons

DESIGN: **CLAUDIA K. LEE**

Paper buttons are great for closures on handmade books, boxes, and envelopes as well as for providing a fun, decorative touch to lampshades, hats, or stationery. These buttons are not made to be used on clothing.

YOU WILL NEED

An assortment of handmade papers of various colors and patterns	Small paintbrush
Scissors	Decorative-edged scissors, or pinking shears
A sheet of heavy handmade paper to use as a background for the collage	Mat knife
White craft glue	Scraps of mat board
Acrylic gloss medium	Small handheld drill or awl
	Scrap of flat, wooden board

1 Cut the pieces of handmade paper into small pieces with scissors. On the sheet of background paper, arrange the small pieces in an interesting, random manner with overlapping edges.

2 Play with the design by moving pieces around until you achieve a composition that you like. Since you'll be cutting this collage into small pieces later, make sure that the compositional 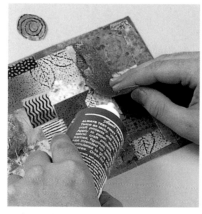 elements are small enough and close enough together to read as interesting compositions after they are cut. Glue the pieces into place with white craft glue, and allow the collage to dry.

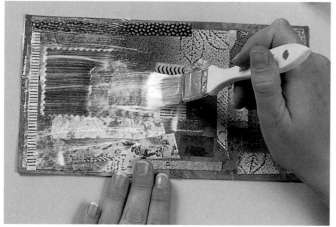

3 Brush both the back and the front of the collage with two coats of acrylic gloss medium. Allow it to dry thoroughly between coats.

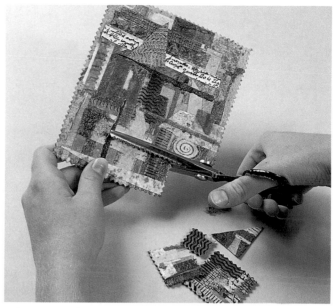

4 Use pinking shears or decorative-edged scissors to cut the buttons into any shape or size that you wish. Use a mat knife to cut out a piece of mat board that is the same shape as the button and slightly smaller. Use white craft glue to attach the mat board to the back of the buttons. Allow the glue to dry.

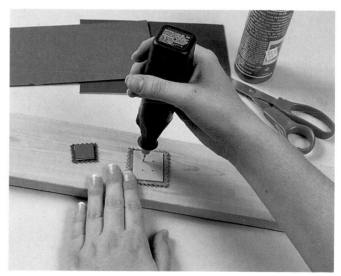

5 Place each of the buttons facedown on the wooden board, and use a small handheld drill or awl to drill or punch holes in each button for attaching them to envelopes, books, or other objects.

Rockin' House Clock

DESIGN: **CLAUDIA K. LEE**

Use the patterns provided for the face of this clock and your own imagination to combine various handmade papers into a fanciful clock.

YOU WILL NEED

Heavyweight paper	White craft glue
Scissors	Gloss acrylic medium
Large sheet of heavyweight handmade paper in assorted colors (we sprayed ours with bright colors)	1-inch-wide (2.5 cm) paintbrush
	Black dimensional paint
	Clock movements
Pencil	Craft knife

1 Use a photocopier to copy the templates on page 75 on heavyweight paper. Use the scissors to cut out the pieces. Use the pencil to trace the patterns onto the handmade paper.

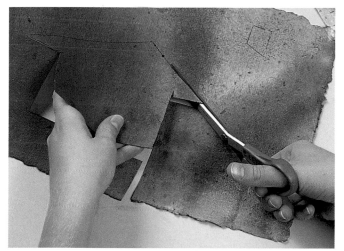

2 Cut out the pieces that make up the house.

3 Using the photo as a guide, position the components of the house. Glue them in place, and allow the piece to dry thoroughly.

4 Use the brush to coat the house with gloss medium, and allow it to dry thoroughly. Apply a second coat, and let dry thoroughly.

5 Using dimensional paint and the photo as a guide, outline the lines of the house with black dimensional paint. Allow the paint to dry thoroughly. Use the craft knife to cut a small "X" in the center of the clock to accommodate the clock movements. Follow the manufacturer's directions for installing them.

Coffeepot Switch Plate Cover

DESIGN: **CLAUDIA K. LEE**

This whimsical switch plate cover will perk up your day every time you turn on your kitchen light.

YOU WILL NEED

Heavyweight paper

Scissors

Pencil

Sheets of handmade paper sprayed with various colors (see project on pages 38 to 39)

White craft glue

Gloss acrylic medium

1-inch (2.5 cm) paintbrush

Single switch plate cover

Scrap of flat wooden board or cutting mat

Craft knife with sharp blade

Flat stick or butter knife

Clear silicone sealant

Awl or ice pick

Black dimensional paint

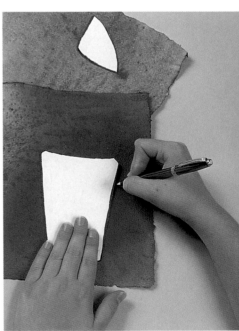

1 Use a photocopier to copy the patterns on page 76 on heavyweight paper. Use scissors to cut out the pattern pieces. Trace the patterns onto various sheets of hand-colored handmade paper. Cut out the shapes.

2 Lay out the shapes in preparation for assembling them. Squeeze glue into the areas marked off by the dotted lines on the edges of the handle and the spout. Press them into place underneath the body of the coffeepot. The rest of the pieces will be glued on top of the coffeepot body, not underneath. Glue the coffeepot's lid into place on top, add the round ball on the crest, and then glue the feet in place at the base of the coffeepot (see the finished picture of the project for guidance).

4 Glue the coffeepot on top of another sheet of handmade paper to serve as the backing. Use scissors to trim away the excess paper. Allow the piece to dry thoroughly.

6 Position the coffeepot faceup on the scrap of board or cutting mat, and use the craft knife to cut out this hole. Then cut away the paper backing that shows inside the crook of the handle, if you wish.

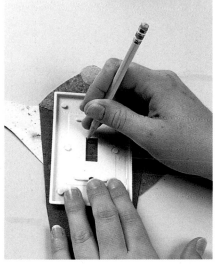

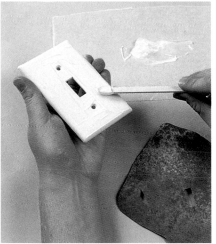

3 Use a small paintbrush to coat the front of the piece with gloss acrylic medium, and allow it to dry thoroughly. Apply a second coat, and allow it to dry thoroughly.

5 Position the face of the switch plate cover on the coffeepot, and use a pencil to trace the small opening for the switch.

7 Use a stick or butter knife to spread a thin coat of clear silicone sealant on the front of the purchased switch plate cover. Align the two switch holes and carefully press the cover into place on the back of the coffeepot. Allow it to dry. Use an awl or ice pick to pierce the holes for the screws using the switch plate cover as a guide. Use black dimensional paint to embellish the surface of the coffeepot by following the outlines of all of the components (see the finished photo for guidance). Add other markings on the surface as you wish.

Guest Book

DESIGN: CLAUDIA K. LEE

This easy-to-assemble booklet can be used as a commemorative guest book for a housewarming or other occasion. To change the theme, all you have to do is change the handmade paper pin that you place on the cover. Use the patterns we've provided on page 77, or make up your own design for the pin.

YOU WILL NEED

5 to 7 sheets of handmade paper for text block, each 12 x 7 inches (30.5 x 17.8 cm)

Bone folder (a bookmaking tool found at art-supply stores)

12 x 7-inch (30.5 x 17.8 cm) piece of heavyweight handmade paper for cover

Ruler

Pencil

Small binder clip or clothespin

Scrap of flat wooden board

Handheld drill or awl and hammer

Size 18 sewing needle

12-inch (30.5 cm) length of waxed linen or cotton crochet cord

Heavyweight paper

White craft glue

Scissors

Colored handmade papers of your choice for pin

Black dimensional paint

3 x 5-inch (7.6 x 12.7 cm) sheet of decorative handmade paper for front of book

Metal pinback

1 Fold each sheet for the text block in half, and crease the folds with the bone folder. Do the same with the cover paper.

2 Lay the book open. Place the ruler along the fold line. Use a pencil to make a notation in the center. Then add marks 2 inches (5.1 cm) on either side of center to make a total of three marks. These marks indicate the sewing stations or holes that will hold the binding thread.

3 Use the binder clip or clothespin to secure one end of the book together so that the pages don't shift. Place the book on a flat scrap of board, and pierce the marked sewing stations with the handheld drill or awl and hammer.

4 Thread the needle with the waxed linen or cotton crochet cord, leaving a short tail and no knots. Bring the needle through the center hole from the outside of the book to the inside. Leave a 3-inch (7.6 cm) tail on the outside of the book. Bring the needle and thread back through either of the other sewing stations to the outside of the book.

5 Bring the needle and thread back through the center again.

6 Bring the needle through the remaining hole to the outside of the book. Slip the needle under the first length of binding stitching, and tie it off with the 3-inch (7.6 cm) tail that you left hanging in step 4. Trim the thread, leaving as much length as you like.

7 Photocopy the pin templates on page 77 onto heavyweight paper. Select one that you want to make into a pin. Follow the same procedure that is outlined in the clock and switch plate projects (pages 50 to 53) to cut out the pieces of the pin, glue the pieces into place (if applicable), and embellish the face of the pin with black dimensional paint. Center and glue the decorative paper to the front cover. Measure the length of the pinback, and use the pencil to mark the placement of both ends of it on the cover.

8 Open out the cover of the book, and place it faceup on the flat board with the pages parted to the other side. Use the handheld drill or awl and hammer to drill or punch holes where the ends of the pinback are located. Glue the pinback onto the back of the pin. Allow the glue to dry thoroughly. Open the pinback, and slide the pin through the holes before closing the clasp. The pin should lie flush with the cover.

Papier-Mâché Vessel

DESIGN: **CLAUDIA K. LEE**

Shape this unusual vessel from handmade paper molded on a balloon.
Make a shorter version of it to create a bowl.

YOU WILL NEED

Jar or cup with a lip that measures approximately 4 inches (10.2 cm) in diameter

Stones or other weighty objects

Balloon inflated to a 9-inch (22.9 cm) diameter

Masking tape

Methylcellulose (a mild glue purchased at art-supply stores or from papermaking suppliers) or wallpaper paste (purchased at home-supply stores)

Freshly pressed sheets of cotton paper with newspaper inclusions (if you're pressing sheets that are 8½ x 11 inches [21.6 x 27.9 cm], you'll need approximately 8 sheets)

1-inch (2.5 cm) brush for glue

1 x 12-inch (2.5 x 30.5 cm) piece of flexible cardboard

12-inch (30.5 cm) length of cord (3 to 4 ply) or yarn for the lip of the vase

Scissors

Sewing needle or straight pin

1-inch-wide (2.5 cm) foam brush

Black acrylic paint

Potted plant (optional)

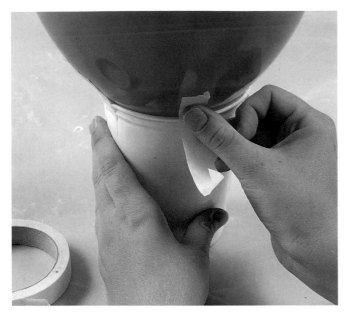

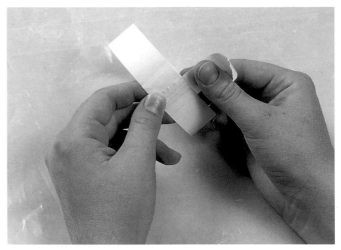

1 Place stones or other objects in the bottom of the jar or cup to anchor it. Place the knotted end of the balloon in the cup, and position it upright. Attach strips of masking tape to the balloon and cup to hold it in place.

3 When you've covered about half of the balloon with papers, make a circle with the strip of cardboard, and tape it in place with masking tape. This ring will serve as the foot of your vase.

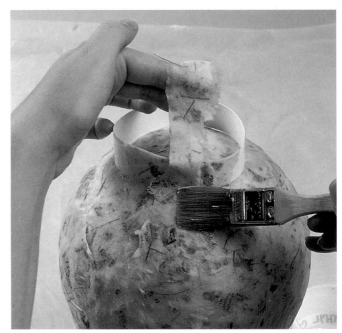

2 If you're using methylcellulose as your glue, combine 16 ounces (480 mL) of water and 8 oz. (224 g) of powder to make a stock solution. If you're using wallpaper paste, mix it according to the package directions. Tear one of the sheets of freshly pressed paper into small strips. Press the strips onto the surface of the balloon, and brush them into place with the glue and the brush. Use the glue generously. Don't worry if you get it on the balloon. To create a strong form, crisscross and overlap the papers as you work.

4 Place the foot at the top of the covered balloon, and center it. Cover it with strips of paper and glue to hold it in place. Add several layers for strength.

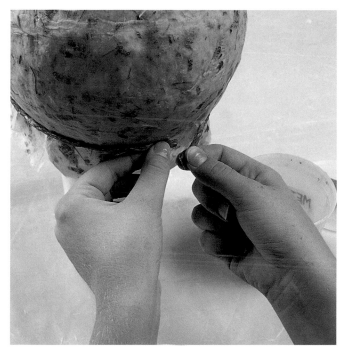

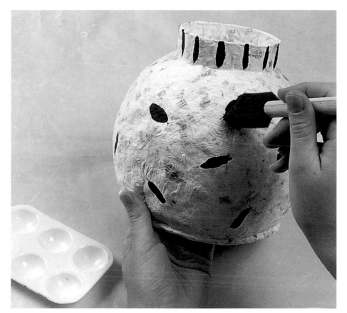

5 Continue covering the balloon. As you near the bottom of it, allow the papers to hang down over the rim of the support container. Use the cord or yarn to circle the vase where the rim will be. Lay it gently in place, and trim it so that the ends just meet rather than overlapping. Remove the cord, and dip it into the glue. Now place it back into position on the balloon.

7 After the form is completely dry, untape the balloon, and turn it over. Pull up on the knotted end of the balloon, and insert a needle or pin to allow the air to escape slowly. Pull the balloon away from the inside of the vase. Use the foam brush and black paint to paint on an overall pattern of shapes to the surface and add stripes to the foot. Allow the paint to dry, and add a potted plant to the vase, if you wish.

SOME OTHER WAYS TO FINISH AND USE YOUR VASE:

1. Sand the surface with fine sandpaper for a smoother finish.

2. Hand stitch colorful threads around the rim or on the surface of the vase.

3. Stitch beads or buttons on the vase.

4. Decorate the surface with collage.

5. Glue on fringe or sew on beads around the rim.

6. Insert a plastic or glass container into the vase, and add cut flowers to it.

7. Use the vase form as a base for a lamp.

8. Make smaller or larger vessels using different-sized balloons.

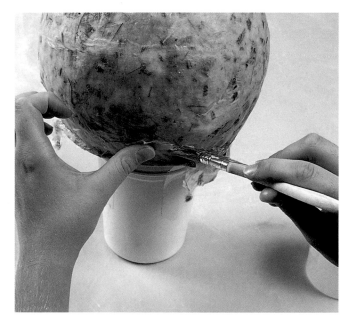

6 Gently sweep the hanging papers up over the cord, and brush them into place with glue. Cover the balloon with a second layer of paper beginning just below the line created with the cord. Allow the form to dry thoroughly.

Deckle-Edged Booklet With Window

DESIGN: **CLAUDIA K. LEE**

Veil a keepsake or treasure in the gauzy window of this handmade paper booklet.

YOU WILL NEED

Rectangular mould and deckle

Duct or masking tape

Decorative item of your choice to enclose in the window such as a leaf, photo, or butterfly wing

Pulp (we used abaca with marigold inclusions, but any pulp is fine)

Plastic vat or tub filled with water

Fabric to use as a felt such as a sheet of wool papermaking felt (see papermaking supplier on page 79), an old wool blanket, cotton sheeting, non-woven interfacing, or muslin

Sponge

Small piece of sheer fabric

Non-fusible interfacing, muslin, or cotton sheeting

Bone folder (bookmaking tool found at art-supply stores)

Scissors (optional)

1 On the topside of the mould and deckle, use the duct or masking tape to mask off about one inch (2.5 cm) of the screen on one of the short ends. Choose the item that you want to place in the window of your booklet. Place this item on the screen of your mould and deckle in a position that leaves about a 1-inch (2.5 cm) margin on one end for binding the book later. Remove the item, and use duct tape to mask off the screen to create a window area large enough to house your item. Prepare your pulp, and add it to the vat of water.

3 Couch the paper on the felt, and press it with the sponge. Remove the mould and deckle to reveal the cover.

2 Use the mould and deckle to pull a sheet of paper. This sheet will serve as the cover of your book. Use your fingers to remove excess pulp that has spilled over onto the taped areas.

4 Cut out a piece of sheer fabric that is at least 1 inch (2.5 cm) larger than the window on all sides (the edges of this piece will be hidden later, so it isn't necessary to cut a perfect shape). Lay it over the window area.

5 Position the decorative item on the sheer fabric. Remove the tape from the center of your mould and deckle, leaving the tape on the edge. Pull a second sheet of paper from the vat. Register, or align, this sheet with the front cover sheet. Couch and press the second sheet onto the first, laminating them together and enclosing your window.

6 Now you'll begin the process of adding pages to the booklet. To begin, cut a piece of interfacing, muslin, or cotton sheeting the same size as the sheet you've couched plus a margin of about an inch (2.5 cm). Cut a piece of this material of the same size for each additional paper sheet that you plan to add to your booklet. Position one of the pieces of interfacing, muslin, or cotton sheeting over the paper, leaving a 1-

inch (2.5 cm) margin exposed on the lower, shorter side of the paper. This area is left exposed for the purpose of binding the pages.

7 Pull another sheet of paper, and align it with the bottom edge of the exposed paper. Couch it on top of the previous sheet in the same position. You'll be laminating the exposed spine as you do this. Continue to laminate pages in this manner until you have added as many as you want. The last sheet that you add will serve as the back cover of your book.

8 Press the book in a press or by hand with a sponge. Allow it to dry thoroughly before removing the fabric from between the pages. Use a bone folder to score a fold line down the spine so that the pages turn nicely. You may also want to trim the spine if it appears too erratic for your taste.

Stamped Box

DESIGN: **EMILY WILSON**

Make a simple box and paper it with stamped or colored handmade papers
that transform both the outside and the inside.

Bookbinder's board or chipboard (available through a craft or art-supply store) cut to the following sizes:
Main panel (bottom and sides):
5½ x 9⅜ inches (14 x 23.8 cm)
Lid: 5½ x 4 inches (14 x 10.2 cm)
Two end panels, each: 2⅜ x 4⅛ inches (6 x 10.5 cm)

Mat knife or hobby knife with fresh blade

Cutting mat or other cutting surface

Ruler

Pencil

Paper scissors

Panels of handmade paper of your choice (ours is decorated with rubber stamped designs) cut to the following sizes:
4 panels (for the lid and bottom), each:
4 x 5½ inches (10.2 x 14 cm)
4 side pieces, each: 2½ x 5½ inches (6.4 x 14 cm)
4 end pieces, each: 2⅜ x 4⅛ inches (6 x 10.5 cm)

Two ½-inch-wide (1.3 cm) brushes

White craft glue

Rubber band

Thin handmade paper (about the consistency of rice paper) cut into 16 strips that measure
½ x 6½ inches (1.3 x 16.5 cm) each

Damp rag

1 Use a mat knife or hobby knife to cut the board into pieces as described in the list above. (This board is extremely dense and will require 6 to10 passes with a mat knife to cut through it completely.) Use the ruler and pencil to draw two parallel lines that are 2½ inches (6.4 cm) from either end of the main panel. Use the mat knife and ruler to score along these lines.

Continue to score and gently remove papers until you've created two ³⁄₁₆-inch-wide (4.8 mm) channels. When cutting the side panels, be careful that you don't cut through to the other side of the paper, since these flaps will be folded up to form sides. Now you're ready to paper the interior of the box. Cut out panels of handmade paper in the sizes listed to the left. Select one of the large paper panels (for the bottom panel of the box), two of the medium-sized panels (for the side panels), and two of the smallest paper panels (for the end panels). Squeeze out a line of glue, and then brush it onto the board (making sure that you cover the edges) before pressing the paper panels into place. Smooth out any air bubbles with your fingertips. Avoid papering into the channels or you'll interfere with the fold of the box.

2 Run a narrow bead of glue along three of the edges of one of the end panels. Fold the main panel into a "U" shape, and position the end piece just inside the edge by holding the unglued edge. Secure the sides for drying by placing a rubber band around the length of the box. Use the extra brush to remove any excess glue from the seams.

3 Allow the glue to set, remove the rubber band, and glue in the other end panel. Secure the box again with the rubber band, and allow the box to dry for 15 to 20 minutes before proceeding.

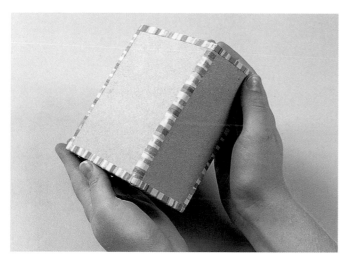

4 Brush glue on the thin handmade paper strips, and apply them one by one to the outside seams of the box, overlapping them and folding them around the edges. As needed, trim the strips with scissors as you work.

5 Leave the top edge of one of the end seams unpapered. (This is where the lid will be attached.)

6 To keep the lid in place while you attach it, cut off a piece of one of the paper strips to a length of about 2 inches (5.1 cm). Fold it in half to double the thickness. Tack half of this 1-inch (2.5 cm) piece of paper length-wise onto the inside of the unfinished edge of the box. Use the other half to attach the lid, so that the paper serves as a temporary hinge. Make sure that the lid will open and close in this position. Close the lid and use one of the strips to paper the outside edge of the lid where it joins the box. (Trim the strip to fit the lid.) Wait for the glue to set on the outside, then open the box and remove the temporary hinge that you placed inside. Use another strip to paper the inside edge of the lid. Smooth down the paper, and then allow the hinge to dry with the box lid propped open. Finish the outside edges of the lid with paper strips.

7 Trim the remaining panel papers by about ⅛ inch (3 mm) on all sides. Brush the papers with glue and apply them to the panels of the box, exposing a portion of the paper edging. Allow them to dry. Apply the bottom panel last, and lay the box on its side to finish drying.

Covered Book with Japanese Stab Binding

DESIGN: **CLAUDIA K. LEE**

If you're intimidated by the thought of making a book, you won't be after making this simple one.

Cover it with sprayed or other handmade papers to suit your taste.

Paper cutter or mat knife and cutting mat	Medium-sized binder clips
10-12 sheets of handmade or commercial papers for pages of the book cut to 5½ x 7½ inches (14 x 19.1 cm) each	Size 18 to 22 tapestry needle with a narrow opening
	Waxed linen or crochet cotton thread in your choice of color
3 sheets of heavy but flexible handmade paper cut to 5½ x 7½ inches (14 x 19.1 cm) each	Sheet of standard-weight paper in color of your choice
Sheet of heavy scrap paper	Scissors
Pencil	1½ x 6-inch (3.8 x 15.2 cm) piece of book board or mat board
Scrap piece of flat wooden board	Bone folder (a bookmaking tool found at art-supply stores) or butter knife
Small handheld drill or awl and hammer	
Hole punch	White craft glue
Eyelet kit	Ribbon or cord

1 Cut a strip of heavy scrap paper that is the length of the book, and use the template on this page as a guide to mark dots with a pencil along the edge of it. (Binding holes can be varied in placement, as shown, but this template gives you the simplest configuration.) Use a pencil to transfer these dots along one of the long edges of one of the cover pieces. Place the dots ½ inch (1.3 cm) in from the edge of the sheet (or what will become the spine's edge). Place the cover on the scrap block of wood, and use a small handheld drill or awl and hammer to drill or tap holes into the cover where the dots are placed.

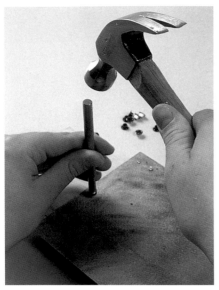

2 Use the hole punch to punch a hole in the top center of the back cover about ½ inch (1.3 cm) from the top. Slip an eyelet into the hole with the front side of it showing on the front of your cover, and use the flange provided in the eyelet kit to tap it into place.

3 Stack the covers and the text block together. Hold them in place with binder clips on the short ends of the book. Follow the marks on the front cover, and drill or punch holes all the way through the book using the handheld drill or awl and hammer.

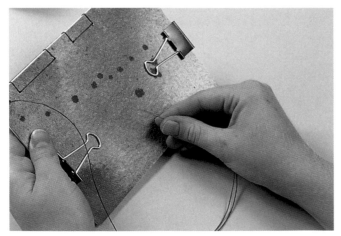

4 Measure out a piece of thread to five times the length of the book. Thread the needle and leave

a 3-inch (7.6 cm) unknotted tail. Follow the instructions on the next page to sew the simplest variation of a stab binding.

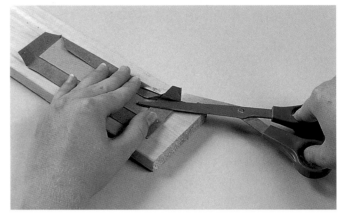

5 From the sheet of standard-weight colored paper, cut out a piece that measures approximately 2½ x 7 inches (6.4 x 17.8 cm). Center the 1½ x 6-inch (3.8 x 15.2 cm) piece of book board or mat board on top of the paper. Fold in the edges and miter them by cutting off the excess with scissors.

6 After you've trimmed the edges of the paper, smear glue on the back of the book board or mat board, press it into place on the paper, and then glue down the mitered flaps. From the third piece of heavy handmade paper, cut a piece that measures 1¼ x 5¾ inches (3.2 x 14.6 cm). Glue this piece of paper into place on top of the mitered edges of the bookmark.

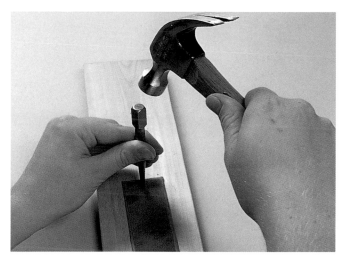

7 Use the handheld drill or awl and hammer to drill or punch a hole in the top center of the bookmark. Cut a 5-inch (12.7 cm) length of ribbon or cord, and fold it in half. Thread the folded edge through the eyelet, and bring the cut ends through the loop. Now thread the cut ends through the hole in the bookmark, and tie an overhand knot to secure the ends. Use a bone folder or butter knife and a ruler to score a fold line close to the binding on the front of the book. Gently fold the book open.

HOW TO SEW A JAPANESE STAB BINDING

1. Thread your needle with waxed linen or crochet thread and leave a 3-inch (7.6 cm) unknotted tail. Beginning at point A, thread the needle through the binding hole to the back of the book. Leave a tail of thread hanging. Bring the thread back around to the front (see right arrow), and insert the needle again. Bring the thread to the back, and adjust it on the front so that it is parallel to the long edge of the book. Now bring the thread back to the front, and insert it in point A again. Pull the thread to the back so that it runs parallel to the short side of the book. Bring the thread to the front at point B.

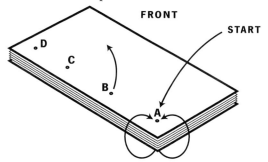

2. Take the thread around to the back and to the front again through the same hole. Now take the thread to point C and through to the back. Bring the thread around to the front, and push it back through point C to the back. Pull the thread up to the front through point D, and then pull it around on the binding side to the back again.

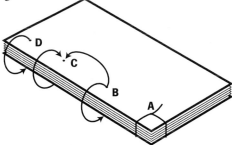

3. Pull the thread back up through point D to the front. Pull it around the front edge of the book and under. Then pull it up through point D again and down through point C to the back. Pull the thread up to the front at point B, and slide it under the threads that cross at point A.

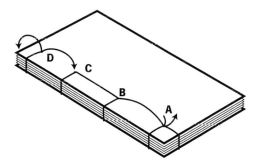

4. Tie off this end with the tail that you left hanging in step 1.

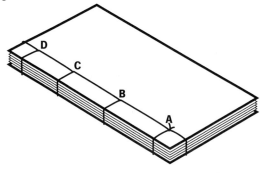

Tabletop Screen

DESIGN: **CLAUDIA K. LEE**

*This unique decorative objet d'art can be used on a table, shelf, or any flat surface.
It's a wonderful way to display and show off beautiful, handmade papers.*

10 strips of 24 x $\frac{1}{16}$-inch (61 cm x 1.6 mm) balsa wood, each cut to 13 inches (33 cm)

Craft saw

Plastic vat or tub

Pulp of your choice

8 x 12-inch (20.3 x 30.5 cm) mould or one that can be masked to that size

Duct or masking tape for masking mould (optional)

Fabric to use as a felt such as a sheet of wool papermaking felt (see paper-making supplier on page 79), an old wool blanket, cotton sheeting, non-woven interfacing, or muslin

Narrow, colored ribbon or string

Sponge

8 x 12 inch (20.3 x 30.5 cm) sheet of fiberglass screening

Piece of mat board or non-corrugated cardboard cut to $1\frac{1}{2}$ x $5\frac{1}{2}$ inches (3.8 x 14 cm)

16-inch (40.6 cm) lengths of colorfast cord or ribbon, approximately $\frac{1}{4}$ inch (6 mm) wide (To test the color, dip a sample of the cord or ribbon into water and blot the cord between two clean paper towels to make certain that it doesn't bleed.)

Drying surface such as sheet of light diffuser grid (plastic), countertop material, glass, or mirror

3-inch (7.6 cm) wide soft bristle brush

1 Use the craft saw to cut the strips of balsa wood to 13 inches (33 cm) each. Fill the vat with water, and add enough pulp to pull a fairly heavy paper. Use the mould to pull the first sheet of paper.

2 Couch the sheet onto the felt. While the mould is still facedown on the felt, lay pieces of colored ribbon or string across the top, bottom, and left side to mark the placement of the mould after you remove it. Press the back of the screen with the sponge. Remove the mould.

4 Pull another sheet, and use the colored ribbons (see step 2) to determine the placement of the mould. Couch this sheet on top of the first, sandwiching the legs and cords between the first and second sheets. Sponge the back of the screen.

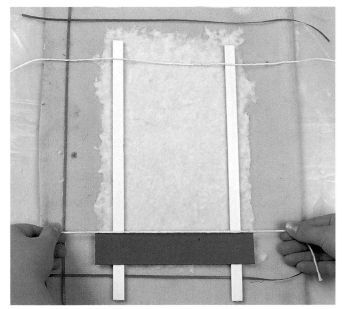

5 Remove the mould to reveal your first screen panel. Place the fiberglass screening over the panel, and press well with the sponge. Flip the felt with the panel on it over onto your drying surface, press the back of it with a sponge to create good contact, and remove the felt. Brush the edges of the paper. Repeat steps 1 through 5 until you have five panels (or more, if you prefer).

3 Place two of the 13-inch (33 cm) balsa strips ½ inch (1.3 cm) from the top of the paper, and 1-inch (2.5 cm) from each of the sides. (These strips will serve as legs for the screen.) With this placement, the extension of the legs will be about 1½ inches (3.8 cm). Make sure that the bottoms of the legs are aligned horizontally. Line up the top edge of the cut piece of mat board or cardboard with the top of the paper. Place one of the lengths of cord along the bottom of this board (see cord pictured at top of photo). Move the board to the bottom of the paper as shown, and place another one of the cords across the top of the board.

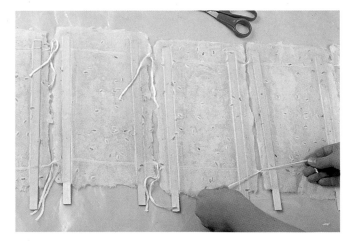

6 After the panels are throughly dry, line up the panels horizontally. Knot the cords together at the top and bottom to attach them closely to one another to form a screen.

Paper Lantern

DESIGN: **JEANNE WHITFIELD BRADY**

This soft, glowing lantern made from sticks and paper is simple to make and gorgeous to look at.

1 Place one sheet of 8 x 17-inch (20.3 x 43.2 cm) handmade paper on a cutting surface. Place the ruler ⅜ inch (9.5 mm) in from the paper's left edge and mark a dotted line with the pencil. Mark four more parallel dotted lines from this starting point that are 4 inches (10.2 cm) apart. You'll end up with a ⅝-inch (1.6 cm) margin on the right side. Along the bottom long side of the paper, measure and mark a line ½-inch (1.3 cm) from the bottom edge. Use the craft knife to cut out random, moon-shaped holes in an overall pattern. Avoid cutting shapes in the ½-inch (1.3 cm) margin at the bottom, or close to the fold where you'll be gluing the legs of the screen. Score the length of each dotted line with the ruler and bone folder.

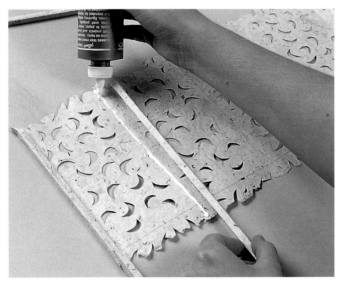

2 Use the craft saw to cut four 11-inch (27.9 cm) long legs from the ½-inch-square (1.3 cm) balsa wood. Use the craft knife to cut four strips from the second sheet of handmade paper that measure 11 x 1¼ inches (27.9 x 3.2 cm) each. Center each balsa leg on one of the strips, spread glue on the inside of the paper, and wrap each leg with the paper so that it is covered. Allow the glue to dry thoroughly. Squeeze a line of glue along four of the five fold lines beginning at the left. Glue the four legs into place along those

lines. (The last fold line along the right edge will remain blank.) Make certain that each of the legs are positioned equidistantly from the top of the paper so that the lantern will be level when it is upright.

3 Cut a piece of ⅛ x ¼-inch (3 x 6 mm) balsa wood to a 4-inch (10.2 cm) length or a length that matches the distance from the bottom of the lantern legs to the straight line along the bottom of the paper on the left end of the lantern. Cut a strip of handmade paper wide enough and long enough to cover this piece (as you did in step 2), and glue it into place. Glue this strip into place on top of the left leg, using the drawn line as a guide for placement.

4 From the ⅛ x ¼-inch (3 x 6 mm) balsa wood, cut four pieces that are each 3⅜ inches (8.6 cm) long. Cut four strips from the handmade paper that measure 1 x 3⅜ inches (2.5 x 8.6 cm) each. Cover the balsa pieces with the strips. Allow them to dry thoroughly. Glue each of the strips into place horizontally along the lower line that marks the lower edge of the lantern. Each strip will fit between the legs.

5 Use the ruler and craft knife to measure and cut mat board to slightly less than 4 inches (10.2 cm) square. Cut out a 2½-inch (6.4 cm) square window from the center of the board. Trim ½ inch (1.3 cm) from each corner. Measure and cut two pieces of handmade paper to the same size as the trimmed square of mat board. Glue one of the pieces of handmade paper to one side of the board. Cut the aluminum screening to 3½ inches (8.9 cm) square, then trim ½ inch (1.3 cm) off the corners.

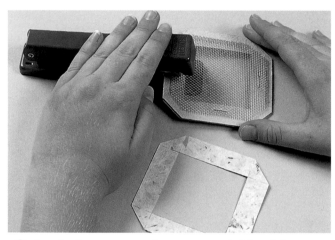

6 Staple the screen to the uncovered side of the mat board. Glue the second piece of handmade paper on top of the screen, covering the raw edges. Stand the lantern up on its legs, and fold the panels into a square configuration. The left edge of the paper that is ⅝ inch (1.6 cm) wide should overlap the last panel. Position the mat board square on the configuration of horizontal strips glued to the bottom of the lantern. This square will serve as a shelf for your candle.

7 Next you'll create a door latch for the lantern from small bits of balsa wood, sticks, or twigs. To do this, use the craft saw to cut one piece of wood to 1¼ inches (3.2 cm) long. Glue it in place

at the centerpoint of the side with the ⅝-inch (1.6 cm) extension edge so that its length protrudes into the adjacent side.

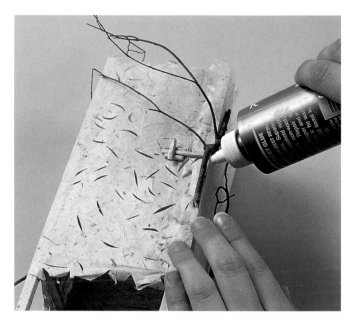

8 Cut three ½-inch-long (1.3 cm) bits of balsa, wood, sticks, or twigs. Glue two of these to the adjacent panel so that they line up on either side of the longer piece. Take the remaining short piece and glue it on top of the two ½-inch (1.3 cm) pieces to hold the longer piece in place. Allow the pieces to dry thoroughly before using the latch. Glue short lengths of decorative twigs to the outside edges of the lantern. Allow everything to dry thoroughly.

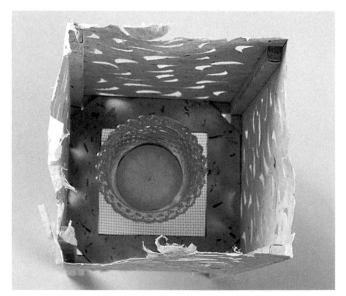

9 Open the door to the lantern, and place the candle and holder inside the lantern on the shelf. Light the candle carefully, and close the latch to the door.

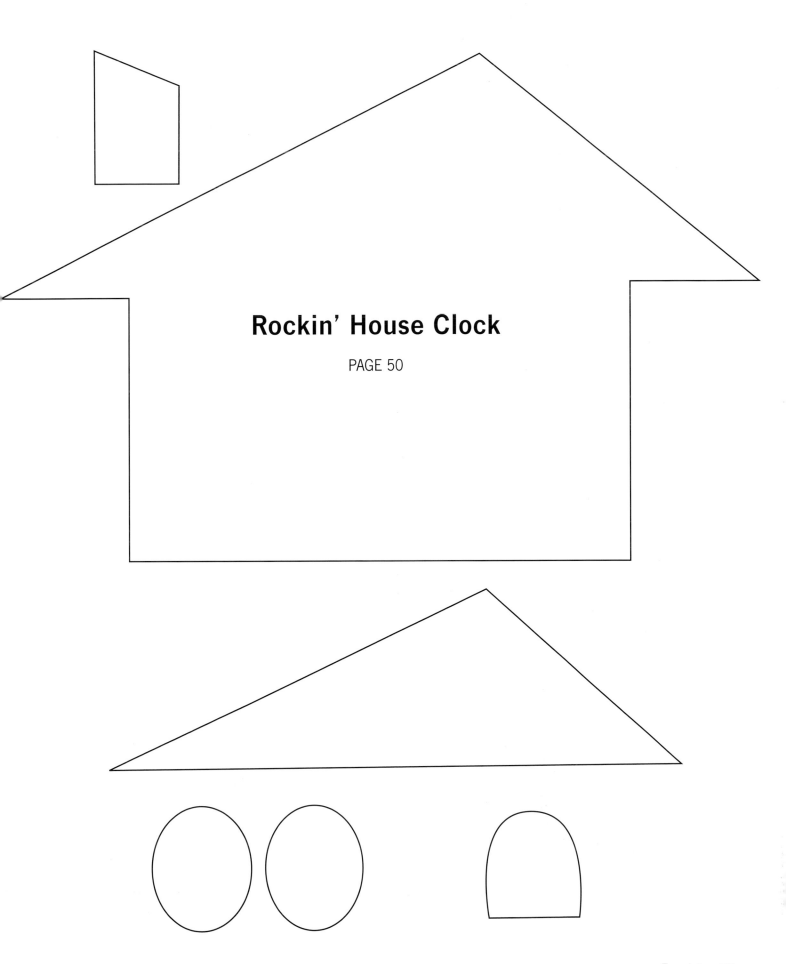

Rockin' House Clock

PAGE 50

Coffeepot Switch Plate Cover

PAGE 52

Guest Book

PAGE 54

DESIGNERS AND CONTRIBUTORS

Jeanne Whitfield Brady is a fiber artist and professor at the Appalachian Center for Crafts in Smithville, Tennessee. Because of her interest in many fiber surfaces and processes, Jeanne divides her creative energies between mixed-media drawing, fabric dyeing with stitched collage, papermaking, and bookmaking.

Arlyn Ende expresses her graphic sensibilities and love of fiber through fabric constructions, mixed-media collages, wall tapetas and rugs for interior architecture, and book design. Her textiles and fiber works are commissioned by corporations, institutions, and private collectors across the United States.

Ann Bellinger Hartley is a mixed-media studio artist living in Houston, Texas. She creates dimensional collages with paper, paint, and found objects. She is the editor of *Bridging Time and Space: Essays on Layered Art* and has work in several books including *Fiberarts Design Books Five and Six* and *Creative Collage* published by Lark Books.

Gail Looper is a bookbinder with a strong interest in decorative papers. She teaches book and paper workshops and is the gallery director at the Appalachian Center for Craft in Smithville, Tennessee.

Beverly Plummer is a papermaker well known for her work with plant pulps and for her vibrant pulp paintings. She had the honor of collaborating with composer John Cage to create macrobiotic papers that he used for a series of prints. She is a new resident of Nashville, Tennessee.

Carlene Taylor is a book artist and workshop instructor who maintains a studio in Cookeville, Tennessee. She received her BFA from the Appalachian Center for Crafts in Smithville, Tennessee.

Emily Tuttle is a printmaker and papermaker who teaches and maintains a studio in Hendersonville, Tennessee. She has a degree in art from the University of Tennessee/Knoxville which prepared her for 25 years of teaching public school art. She has worked for ten years as an independent craft artist.

Sandy Webster is a mixed-media artist who lives in North Carolina. Her works have appeared in many publications and juried exhibitions throughout the United States. She teaches in the United States, Canada, and Australia.

Emily Wilson is a sculptor living in Middle Tennessee with her husband, Tim Hintz, who is a green wood chairmaker.

GLOSSARY

ADDITIVES Ingredients in paper, other than pulp, such as pigments, dyes, sizing, or retention aid.

CELLULOSE The main part of the cell wall of a plant necessary for making paper.

COUCH The act of transferring a newly formed sheet of paper from the mould to the felt.

DECKLE A frame that sits on top of the mould to contain the pulp when forming sheets of paper. The use of a deckle results in a harder edge and is useful when making thicker sheets or using up smaller amounts of pulp.

DECKLE EDGE The feathery edge of paper formed by using the mould without the deckle.

EMBOSS To impress a design, pattern, or image in a sheet of paper by pressing a three-dimensional object into the surface of the paper by hand or with the aid of a press.

FELT An absorbent natural or synthetic fabric used to separate sheets of newly formed paper as they are pressed.

INCLUSIONS Materials added to pulp such as flower petals, confetti, dirt, or leaves. Inclusions can be added directly to the vat or toward the end of the processing cycle.

KISS-OFF To return a newly formed sheet that is still on the mould to the vat by turning the mould upside down over the vat and gently touching the surface of the water to release the paper.

LAMINATE To couch a newly formed sheet of paper onto another sheet instead of onto a felt.

LIGNIN A natural glue-like substance that holds together the cellulose fibers in some plants. If left in pulp, it causes the paper to yellow and become brittle. Cooking plant fibers in a caustic solution helps remove lignin.

PIGMENT Finely ground particles added to pulp with a retention aid to color the pulp.

POST A stack of wet sheets of paper on felts that are placed on a press to remove more water from them.

PULP Plant fibers that have been processed for papermaking. Papermaking pulps include those derived from plants that have been cooked and beaten as well as purchased sheets and recycled papers.

RETENTION AID A chemical that is added to pulp to create a bond between pigments or dyes and the fibers.

SIZING A chemical that is added to paper in the pulp stage or applied to the finished sheets that causes them to be less absorbent.

SLURRY A mixture of pulp and water.

VAT A tub or container that holds the water and pulp for the purpose of dipping.

PAPERMAKING SUPPLIERS

If you choose to use wool papermaking felt (as listed in many of the projects), and can't find it at a local art supply store, you can order it from a papermaking supplier. Sample companies that sell this product are listed below.

Carriage House Papers
79 Guernsey Street
Brooklyn, NY 11222
Phone/fax: 718-599-7857
E-mail: chpaper@aol.com

Lee Scott McDonald, Inc.
P.O. Box 264
Charlestown, MA 02129
Phone: 1-800-627-2737
Fax: 617-242-8825

Twinrocker Handmade Paper
P.O. Box 413
Brookston, IN 47923
Phone: 1-800-757-8946
E-mail:twinrock@twinrocker.com
www.twinrocker.com

The Papertrail
1546 Chatelain Avenue
Ottawa, Ontario K12885
Canada
Phone: 613-728-4669
Fax: 613-728-7796

Paper Capers
P.O. Box 281
Neutral Bay, N.S.W. 2089
Australia
Phone/fax: 61-29964-9471

Pulp and Paper Information Center
Papermaker's House
Rivenhall Road
Swindon, Wilts SN5 7BE
England
Phone: 44 (0) 1793 886086
Fax: 44 (0)1793 886182

INDEX